Cambridge Elements ≡

Elements of Paleontology
edited by
Colin D. Sumrall
University of Tennessee

VIRTUAL PALEONTOLOGY

Tomographic Techniques for Studying Fossil Echinoderms

Jennifer E. Bauer
University of Michigan Museum of Paleontology

Imran A. Rahman
Natural History Museum and Oxford University Museum of Natural History

CAMBRIDGE
UNIVERSITY PRESS

University Printing House, Cambridge CB2 8BS, United Kingdom

One Liberty Plaza, 20th Floor, New York, NY 10006, USA

477 Williamstown Road, Port Melbourne, VIC 3207, Australia

314–321, 3rd Floor, Plot 3, Splendor Forum, Jasola District Centre,
New Delhi – 110025, India

103 Penang Road, #05–06/07, Visioncrest Commercial, Singapore 238467

Cambridge University Press is part of the University of Cambridge.

It furthers the University's mission by disseminating knowledge in the pursuit of
education, learning, and research at the highest international levels of excellence.

www.cambridge.org
Information on this title: www.cambridge.org/9781108794763
DOI: 10.1017/9781108881944

First published 2021

A catalogue record for this publication is available from the British Library.

ISBN 978-1-108-79476-3 Paperback
ISSN 2517-780X (online)
ISSN 2517-7796 (print)

Additional resources for this publication at cambridge.org/virtualpaleontology.

Virtual Paleontology

Tomographic Techniques for Studying Fossil Echinoderms

Elements of Paleontology

DOI: 10.1017/9781108881944
First published online: September 2021

Jennifer E. Bauer
University of Michigan Museum of Paleontology

Imran A. Rahman
Natural History Museum and Oxford University Museum of Natural History

Author for correspondence: Jennifer E. Bauer, bauerjen@umich.edu

Abstract: Imaging and visualizing fossils in three dimensions with tomography is a powerful approach in paleontology. Here, the authors introduce select destructive and non-destructive tomographic techniques that are routinely applied to fossils and review how this work has improved our understanding of the anatomy, function, taphonomy, and phylogeny of fossil echinoderms. Building on this, this Element discusses how new imaging and computational methods have great promise for addressing long-standing paleobiological questions. Future efforts to improve the accessibility of the data underlying this work will be key for realizing the potential of this virtual world of paleontology.

Keywords: tomography, echinoderms, digital data, virtual paleontology, 3D visualization

ISBNs: 9781108794763 (PB), 9781108881944 (OC)
ISSNs: 2517-780X (online), 2517-7796 (print)

Contents

1 Introduction

The availability of increasingly powerful computers has transformed the ways in which paleontologists study fossils, opening up a plethora of opportunities for processing, analyzing, and simulating paleontological data. This has included the development of an entirely new approach, termed virtual paleontology (Sutton et al. 2014, 2017), which entails using modern computational methods to generate three-dimensional digital representations of fossil specimens. These "virtual fossils" have enabled the characterization and reassessment of form, function, and phylogeny in greater detail than previously possible, and the techniques involved are broadly applicable across a wide range of taxonomic groups, specimen sizes, and modes of preservation (Cunningham et al. 2014; Rahman & Smith 2014; Sutton et al. 2014, 2017).

The study of 3D structures by creating a series of 2D parallel "slices" through an object, or tomography, enables exploration of internal anatomy and gross morphology in different ways. Tomographic techniques are among the most powerful methods in virtual paleontology. Traditional methods relied on creating physical sections through fossils using serial grinding or sawing, with images captured by manual tracing or acetate peels (e.g. Sollas 1904, Stensiö 1927) and more recently digital photography (e.g. Sutton et al. 2001); although potentially informative, these approaches are laborious and destructive, restricting their utility for studying fossil samples (Cunningham et al. 2014). However, in the past two decades, non-destructive X-ray computed tomography (CT) and its variants (e.g. micro-CT and synchrotron tomography) have rapidly grown in prominence, due in part to the increasing accessibility of suitable facilities (Sutton et al. 2014, 2017). This has driven the rise of tomography in paleontology.

Vertebrates were an initial focus of study in virtual paleontology (e.g. Tate & Cann 1982; Conroy & Vannier 1984; Haubitz et al. 1988; Rowe et al. 2001), but in recent years invertebrate fossils have been the subject of a growing number of tomographic investigations. Echinoderms are an invertebrate group well-suited to these techniques; they have an excellent fossil record, comprised of numerous three-dimensionally preserved specimens that are amenable to tomography. Moreover, the phylum represents a diverse, disparate, and geographically and temporally widespread clade, providing an ideal model system for addressing complex ecological and evolutionary questions through time (e.g. Sheffield et al. 2018; Cole et al. 2019; Deline et al. 2020). Research using tomographic techniques has addressed aspects of echinoderm morphology, function, preservation, and phylogeny with three-dimensional virtual fossils (e.g. Stock & Veis 2003; Sutton et al. 2005; Schmidtling & Marshall 2010; Rahman et al. 2015). These studies have led to a better understanding of, for example, homology, informing on evolutionary relationships and transitions among major echino-derm groups.

Herein, we summarize advances in understanding the anatomy, function, taphonomy, and evolutionary history of extinct echinoderms obtained through the generation and study of virtual fossils. In addition, we review the methods used in this work and discuss possible future developments with particular relevance to echinoderm paleobiology. We focus on the applications of tomographic techniques, especially physical–optical and X-ray computed tomography; surface-based methods (e.g. laser scanning and photogrammetry) also have great potential for studying fossil echinoderms (e.g. Hendricks et al. 2015), but are not detailed herein.

2 Methods

Tomographic techniques have a long history of use in paleontology, dating back more than a century (e.g. Sollas 1904), but it is only in the last 10–20 years that they have become a routine part of the paleontologist's tool kit. In the following section, we introduce physical–optical and X-ray tomography, which are the two methods most commonly applied to fossil echinoderms. For more detailed treatments of these and other tomographic techniques, with a focus on their applications in paleontology, see Sutton et al. (2014, 2017).

2.1 Physical–Optical Tomography

This method involves physically exposing surfaces of a fossil specimen, which are then optically imaged to create a dataset of slice images (Figure 1, parts 1–3). Surfaces are typically exposed by serial grinding, with the exposed surfaces captured at regular intervals through manual tracing, acetate peels, or photography (Sutton et al. 2014, 2017). These two-dimensional slices can be used to generate physical or digital models of the specimens, enabling three-dimensional visualization of fossil morphology (virtual fossils; Sollas 1904; Stensiö 1927; Schmidtling and Marshall 2010; Huynh et al. 2015). The major drawbacks of this process are that it is at least partially destructive and very time consuming, although the speed of imaging has been greatly improved by the integration of digital photography (Sutton et al. 2014, 2017). Physical–optical tomography remains an especially useful approach for studying fossil specimens that cannot be physically or chemically prepared out of the host rock and which do not provide sufficient internal density contrast for X-ray imaging techniques (e.g. some calcite fossils preserved in calcareous sediments).

Physical–optical tomography has been extensively used to study fossil invertebrates from the Herefordshire Lagerstätte of England (Briggs et al. 1996;

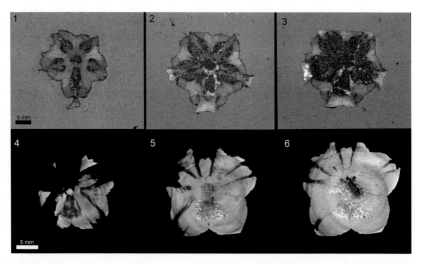

Figure 1 Comparison of destructive and non-destructive tomography applied to
the blastoid *Cryptoschisma*. (1–3) Acetate peels generated by A. Breimer and
digitized by J. A. Waters, housed physically and digitally at the Naturalis
Biodiversity Center RGM.986207. (4–6) Synchrotron X-ray tomographic
slices. Slices are available in the Supplemental Material and the specimen is
housed at Museo Geominero (MGM), Instituto Geológico y Minero de España
(MGM-3384D).

Sutton et al. 2001; Siveter et al. 2020), including echinoderms (e.g. Sutton et al.
2005; Briggs et al. 2017; Rahman et al. 2019). This deposit is early Silurian in
age (Wenlock Series, ca. 425 Ma) and preserves fossils three-dimensionally as
calcite infills in calcareous nodules, sometimes with details of soft tissue
anatomy (Orr et al. 2000; Sutton et al. 2001, 2005; Siveter et al. 2020).
Because the fossils are typically very similar in chemical composition to the
host rock, non-destructive X-ray tomography has generally not proved effective
(although see Nadhira et al. 2019). Consequently, specimens have been studied
by necessity using destructive techniques. Sutton et al. (2001) developed
a protocol for serially grinding and digitally photographing specimens, which
has subsequently been applied to an assortment of Herefordshire fossils (see
Siveter et al. 2020 and references therein).

2.2 X-Ray Tomography

The most widely used tomographic techniques in paleontology rely on X-rays
for non-destructive imaging of fossil samples (i.e. X-ray computed tomography
or CT). These methods trace their roots back to the end of the nineteenth
century, when X-rays were first used to capture images of fossils (Branca

1906). CT involves penetrating the specimen of interest with an X-ray beam and capturing a set of radiographic images at multiple angles, which are then used to reconstruct a tomographic dataset that maps X-ray attenuation within the specimen (Sutton et al. 2014, 2017). Such datasets can be visualized in three dimensions using a multitude of free (e.g. Drishti, SPIERS) and commercial (e.g. Avizo, Mimics, VG Studio Max) software packages (see Cunningham et al. 2014 and Sutton et al. 2014 for more detailed treatments), thereby virtually extracting the fossil from the surrounding rock. While medical CT scanners have proved useful for studying some large vertebrates, laboratory-based, high-resolution variants (i.e. micro-CT and nano-CT) are better suited for most fossil specimens (e.g. University of Texas High-Resolution X-ray Computed Tomography Facility, UTCT). Micro-CT, in particular, has broad utility in paleontology; it can be used to penetrate dense samples while at the same time resolving fine anatomical details for a range of specimen types. Additionally, micro-CT equipment is now more widely available and accessible to many researchers.

Micro-CT scanners can typically achieve much higher resolutions than medical systems, down to a few microns or less, and are not restricted by the need to minimize radiation dosage (as is the case for medical CT) (Sutton et al. 2014, 2017). Settings such as the energy of the source X-rays can be adjusted to provide a balance between penetration and contrast (according to the composition of the sample). The need to minimize artifacts, which can obscure anatomical details, is another consideration when determining the optimal scan settings. Beam hardening is an especially common artifact when scanning large and dense fossils with micro-CT and has the effect of artificially darkening the center of the slices; this can be partly addressed through the use of filters, thin pieces of metal that are positioned between the source and sample to increase the average X-ray energy (Sutton et al. 2014, 2017).

Conventional CT techniques require that the fossil and matrix be composed of different materials, as chemically homogenous samples do not normally provide sufficient X-ray attenuation contrast for imaging in this way (Sutton et al. 2014, 2017). A common example of this problem is calcitic fossils embedded in calcareous rock, which can be very difficult to differentiate with lab-based CT scanners. Synchrotron tomography provides a potential solution to this problem (Figure 1, parts 4–6). This method uses a particle accelerator to generate extremely bright X-rays, producing beams that consist of X-rays of a single energy (i.e. monochromatic X-rays). Synchrotron tomography beamlines are optimized for different sample sizes and resolutions, and are capable of enhancing contrast between similar materials through phase-contrast imaging, which exploits X-ray refraction at material interfaces (Cunningham et al. 2014;

Sutton et al. 2014). Some lab-based nano-CT scanners can be used to improve contrast in a similar way (e.g. Dunlop et al. 2012).

3 Anatomy and Function

For many groups of fossil echinoderms, the most detailed descriptive studies remain those published in the mid- to late-twentieth century, particularly the *Treatise on Invertebrate Paleontology* and a handful of other seminal works (e.g. Moore et al. 1952; Sprinkle 1973). However, this work was frequently restricted to the exposed surfaces of fossil specimens, with key details hidden in the surrounding rock. Moreover, internal anatomy proved difficult to describe using traditional methods. Serial sectioning of specimens was undertaken in some cases (e.g. Breimer & Macurda 1972) but was typically incomplete due to the internal structure of interest (e.g. gonads or respiratory structure) terminating within the sample, the time required for such work, and the difficulty in visualizing the results, which were often hand-drawn or traced from the sections. Consequently, virtual paleontology has much potential for improving knowledge of both the external and internal anatomy of extinct echinoderms, with implications for functional morphology, and this has been a growing area of research in echinoderm paleobiology over the past two decades.

3.1 External Morphology

3.1.1 Carpoidea

Carpoids are an enigmatic and controversial extinct group, most likely paraphyletic or polyphyletic, that includes ctenocystoids, stylophorans, solutes, and cinctans (David et al. 2000; Smith 2005). Representatives of the group are weakly to strongly asymmetrical or bilaterally symmetrical, unlike many modern echinoderms, which exhibit pentaradial symmetry as adults (Zamora & Rahman 2014). Commonly, specimens are three-dimensionally preserved as molds with no original skeletal material remaining. Traditionally, latex casts have been prepared from such fossils, allowing surface anatomical features to be described, but this process risks damaging fragile specimens. An alternative approach is to digitally reconstruct fossils non-destructively using X-ray computed tomography. One example of such work is the study of the Cambrian echinoderm *Ctenoimbricata spinosa* Zamora et al. 2012. Two specimens were collected, both preserved three-dimensionally as molds with part and counterpart. Micro-CT was used to image and digitally reconstruct these fossil specimens. This allowed *C. spinosa* to be virtually extracted from the rock surrounding it, described as a new genus and species, and used as a basis for inferring homologies with ctenocystoids and cinctans (Figure 2). Rahman et al.

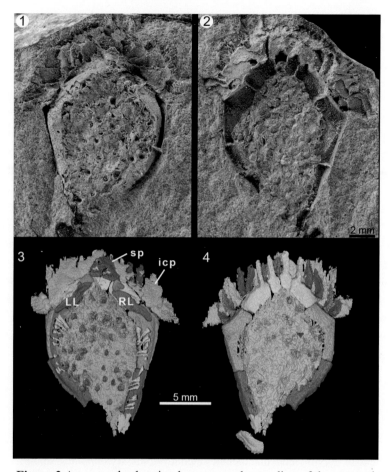

Figure 2 An example showing how our understanding of the external
morphology of fossil echinoderms has been improved through the application
of X-ray computed tomography. Specimens preserved as part and counterpart
can be digitally extracted from the surrounding matrix to produce a complete
three-dimensional model. (1–2) Natural mold of *Ctenoimbricata spinosa* (MPZ
2011/93), (1) dorsal and (2) ventral views. (3–4) Virtual reconstructions of the
same specimen in (3) dorsal and (4) ventral views. Modified from Zamora et al.
(2012).

(2015) used micro-CT to image a slab of siltstone preserving multiple moldic
specimens of the Ordovician ctenocystoid *Conollia sporranoides* Rahman et al.
2015. This revealed a number of fossils that were previously hidden within the
rock, providing anatomical details that informed a description of the new
species and a redescription of the genus. Lastly, Rahman et al. (2010) described
the Cambrian stylophoran *Ceratocystis prosthiakida* Rahman et al. 2010,

preserved as a mold in shale, with the aid of micro-CT. Using the resulting virtual fossil, they were able to digitally restore the original orientations and articulations of thecal plates, thereby providing a more accurate three-dimensional model of the animal's external morphology and helping to identify it as a new species.

More rarely, carpoid fossils are preserved three-dimensionally as recrystallized calcite, embedded in matrix. Historically, this material was studied through serial sectioning (e.g. Jefferies & Lewis 1978). However, such specimens are also well suited for micro-CT, as long as there is sufficient density contrast between fossil and rock. Domínguez et al. (2002) were among the first to apply micro-CT to fossil echinoderms, digitally reconstructing a three-dimensional specimen of the Carboniferous stylophoran *Jaekelocarpus oklahomensis* Kolata et al. 1991 and redescribing its anatomy. The oldest cinctan yet discovered, *Protocinctus mansillaensis* Rahman and Zamora 2009 from the Cambrian of Spain, was described as a new genus and species based on micro-CT scans of three specimens preserved as recrystallized calcite in siltstone (Rahman & Zamora 2009). The resulting digital reconstructions clearly showed the posterior appendage in an oblique orientation, suggesting it functioned as an anchor in the living animal. Rahman and Clausen (2009) used micro-CT to image a specimen of the Cambrian ctenocystoid *Ctenocystis utahensis* Robison and Sprinkle 1969, which is preserved three-dimensionally as a mold coated in recrystallized calcite. This allowed the species to be described in greater detail than previously possible, shedding new light on the morphology and function of key ctenocystoid characters, such as the anterior ctenoid apparatus (a series of distinctive bladelike plates), a structure considered to be used for feeding and/or respiration.

3.1.2 Blastozoa

Blastozoan anatomy was thoroughly described by Sprinkle (1973). Much of this work was restricted to the exposed surfaces of fossil specimens, augmented with serial sections revealing internal details (Sprinkle 1973). As with carpoids, three-dimensionally preserved fossil specimens are known, many of which are potentially amenable to investigation using tomographic techniques. However, a major challenge is differentiating between fossils preserved as calcite and the infilling sedimentary matrix, often rich in calcium carbonate, which can provide insufficient internal density contrast for conventional micro-CT. In such cases, synchrotron tomography can be effective (see Section 2.2). Bauer et al. (2019) applied this technique to two specimens of the oldest recorded blastoid, *Macurdablastus uniplicatus* Broadhead 1984 (Late Ordovician of eastern Tennessee). They scanned specimens using propagation-based phase-contrast synchrotron tomography on the TOMCAT beamline at the Swiss Light Source. This allowed them to better resolve the boundaries between plates in the oral area and thereby identify homologous structures shared with other blastozoans (including coronoids and other conservatively plated species; see Section 5 on phylogeny).

If fossil specimens are moldic, micro-CT is often the method of choice because suitable instruments are widely available (synchrotron tomography relies on access to a small number of paleontologically suitable beamlines). Zamora and Smith (2012) used this technique to image two specimens of the Cambrian blastozoan *Dibrachicystis purujoensis* Zamora and Smith 2012. This helped them elucidate the morphology of the paired feeding appendages, indicating that these structures arose as thecal extensions of the body, with plating changing from proximal to distal. These were interpreted as key characters and supported a description of the fossils as a new genus and species.

3.1.3 Crinoidea

Combining traditional paleontological methods and non-destructive micro-CT, Zamora et al. (2015) thoroughly described a new species of iocrinid crinoid, *Iocrinus africanus* Zamora et al. 2015 (Middle Ordovician of Morocco). This specimen is preserved as a mold with no remaining skeletal material. Micro-CT imaging enabled the digital reconstruction of the specimen in three dimensions, bringing to light key anatomical features of the column, distal arms, and aboral cup that were recorded as impressions in matrix. These details allowed Zamora et al. (2015) to properly assign the material to *Iocrinus* and recognize it as a new species.

3.1.4 Cyclocystoidea

The anatomy of a new cyclocystoid genus and species, *Moroccodiscus smithi* Reich et al. 2017, from the Middle Ordovician of Morocco, was described with the aid of micro-CT scanning (Reich et al. 2017). Specimens are preserved three-dimensionally as part and counterpart molds in concretions. Details gained from the virtual reconstructions of *Moroccodiscus*, including the round flattened shape, domed central disk, and lack of an enclosed body cavity, suggest that cyclocystoids most likely lived with their flat marginal surface down, resting on or attached to the ocean floor (Reich et al. 2017).

3.1.5 Ophiuroidea

Many Paleozoic ophiuroids, or brittle stars, possessed anatomy different from that of their modern counterparts, which makes inferences of life mode difficult. Clark et al. (2020a) presented six virtual ophiuroid fossils from the Lower

Devonian Hunsrück Slate, with some taxa displaying typical Paleozoic arm structures and others with arm morphologies more similar to modern taxa. The use of computed tomography promoted the visualization of the delicate arms in situ and allowed for detailed descriptions of the arms. This work allowed for the inference of different locomotion strategies in ophiuroids.

3.2 Respiratory and Water Vascular System

The water vascular system (WVS) is a network of fluid-filled coelomic canals that is unique to Echinodermata (Nichols 1972; Zamora & Rahman 2014). This system has been hypothesized to perform a variety of functions in fossil echinoderms, including feeding, locomotion, and respiration. Although typically inferred in fossil specimens indirectly based on skeletal features such as the hydropore and surrounding plates (e.g. Dean 1999; Sprinkle & Wilbur 2005), in some groups direct evidence of the water vascular system is preserved in the form of calcified structures within a tightly sutured theca (e.g. Paul 1967, 1968; Sprinkle 1973; Sumrall & Waters 2012). Even more rarely, soft parts of the water vascular system, such as the tube feet, can be preserved in fossil specimens (e.g. Glass 2006; Ausich et al. 2013; Lefebvre et al. 2019); these parts have typically been identified through elemental analyses and scanning electron microscopy. Below, we summarize key advances in understanding the water vascular system enabled by virtual paleontology.

3.2.1 Eublastoidea

Traditionally, characteristics of the respiratory system of blastozoan echinoderms have been taken as synapomorphies for defining taxonomic groups and delineating species (Sprinkle 1973; Bauer et al. 2017; see Sheffield et al. forthcoming in this series for a full discussion on this topic). Endothecal, or internal, respiratory structures of blastozoans exist as infoldings of the body wall, and as lightly calcified structures they are often preserved when thecae are intact (Paul 1968; Sprinkle 1973; Sumrall & Waters 2012). These structures have been examined and described by serially sectioning specimens (e.g. Beaver et al. 1967; Breimer 1988a, 1988b; Dexter et al. 2009; Schmidtling & Marshall 2010). Recent efforts have focused on digitally reconstructing eublastoid hydrospire structures from serial sections (e.g. Schmidtling & Marshall 2010; Waters et al. 2015, 2017; Huynh et al. 2015; Bauer et al. 2017).

Schmidtling and Marshall (2010) provided the first attempt to create three-dimensional hydrospires from serial sections of *Pentremites rusticus* Hambach 1903, an early Pennsylvanian species from Oklahoma, USA. The specimen was serially sectioned across several sessions at regular intervals, and the hydrospire groups of ambulacra C and B were digitally reconstructed. This confirmed that these two hydrospire groups merged to form a single spiracle. This work also included an analysis of how fluid was able to flow through the hydrospires, and helped show the amount of space these structures occupy within the theca (Schmidtling & Marshall 2010). The work of Schmidtling and Marshall (2010) spurred the efforts of Huynh et al. (2015), who examined fluid flow through the same reconstructed hydrospires to test hypotheses about the function of the hydrospires (see Sheffield et al. forthcoming in this series).

Waters et al. (2015) were the first to digitally reconstruct complete eublastoid hydrospires in three dimensions, using a serial acetate peel dataset produced by Breimer in the 1960s (e.g. Breimer & Macurda 1972), thus providing an incredibly detailed look at these anatomically complex structures (Figure 3, parts 1–2). This work served as a baseline for reconstructing hydrospires in different species spanning Eublastoidea, informing descriptions of the morphology as a whole rather than partial descriptions based on individual sections (e.g. Beaver et al. 1967; Breimer 1988a, 1988b). For example, Bauer et al. (2017) created virtual models of the hydrospires of six eublastoid species from the Carboniferous–Permian of the United States and Timor by digitizing a dataset of serial acetate peels. This allowed them to redescribe the anatomy of the hydrospires in great detail, identifying differences within and between previously recognized families.

The increasing availability of non-destructive imaging techniques, such as X-ray tomography, has provided new opportunities to reexamine the internal structures of blastozoans. This proved particularly important for the Late Ordovician blastoid *Macurdablastus*, which was previously unassignable outside of class Blastoidea. Only a few well-preserved specimens are known, therefore traditional destructive techniques such as serial sectioning could not be used. The two most complete specimens (holotype and paratype) were analyzed using synchrotron tomography in order to enhance the contrast between fossil and matrix (Bauer et al. 2019). This revealed that the morphology of the reconstructed respiratory structures differs from that of the two known eublastoid orders, fissiculates and spiraculates. See Sheffield et al. forthcoming (in this series) for a more detailed overview of blastozoan water vascular systems.

3.2.2 Asterozoa

In extant asterozoans, the WVS consists of a ring canal and a set of radial canals, each of which gives rise to numerous lateral canals and tube feet. Among fossil forms, aspects of this system can be inferred indirectly from skeletal

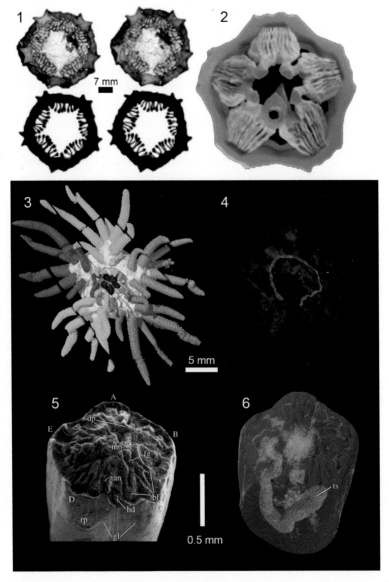

Figure 3 Examples of advances in understanding the internal anatomy of various extinct echinoderm groups. (1–2) Internal anatomy of *Pentremites godoni*, modified from Waters et al. (2015). (1) Digitized acetate peels with sketched internal anatomy and plate thickness. (2) Ceramic model of the reconstructed respiratory structures. (3–4) Reconstruction of *Sollasina cthulhu* (OUMNH C.29662) modified from Rahman et al. (2019). (3) Oral view of the reconstruction. (4) Transparent oral view showing an internal ring-like structure. (5–6) Post-metamorphic blastoid specimen, modified from Rahman et al. (2015). (5) Oblique view of the top of the specimen. (6) Same view but of a transparent reconstruction to showcase the U-shaped gut.

morphology (e.g. Dean 1999); under exceptional circumstances, these soft tissues are preserved in the fossil record (e.g. Glass 2006), providing direct insights into the nature and function of the WVS. Clark et al. (2017) used micro-CT to visualize the water vascular system in an extraordinarily well-preserved specimen of the Ordovician ophiuroid *Protasterina flexuosa* (Miller & Dyer 1878) with pyritized soft parts (Glass 2006). The resulting digital reconstruction provided a very detailed, three-dimensional view of the anatomy of the preserved WVS, including the radial canals, lateral branches, and tube feet. This suggested that the radial canals did not possess sphincters for controlling the turgidity of tube feet, as seen in extant ophiuroids, indicating that the WVS of *Protasterina* functioned differently from that of living species.

Sutton et al. (2005) described new asterozoan material from the lower Silurian Herefordshire Lagerstätte of England. Like other fossils from this site, specimens are preserved three-dimensionally as calcite infills in nodules (see Section 2.1). To better visualize the morphology of this material, specimens were serially ground and imaged with digital photography to create virtual representations of the fossils. These virtual fossils revealed the presence of a pyloric system in the gut and large bivalved pedicellariae, as well as elements of the water vascular system including ampullae, podia, and a radial canal (Sutton et al. 2005). These details align *Bdellacoma* Salter 1857 more closely with Asteroidea than with Ophiuroidea, as previously suggested. Details of the tube feet and other aspects of soft-tissue anatomy suggest *Bdellacoma* likely belongs within the asteroid crown group (Sutton et al. 2005).

3.2.3 Edrioasteroidea

Edrioasteroids are an extinct group of echinoderms characterized by a disklike to globular theca with five ambulacra. Abundant three-dimensionally preserved fossils are known, and the study of this material has enabled detailed descriptions of skeletal morphology. This has been used as a basis for reconstructing the edrioasteroid WVS (e.g. Bell 1977; Paul & Smith 1984; Smith 1985), but direct evidence of soft parts was lacking. Briggs et al. (2017) described a new genus and species of edrioasteroid, *Heropyrgus disterminus* Briggs et al. 2017, from the Herefordshire Lagerstätte. They selected three specimens for serial grinding, and the resulting digital reconstructions revealed soft-tissue preservation in one of these specimens. Soft parts consist of tube feet arranged into upper and lower sets, attaching to the inner surfaces of the compound interradial plates. Other probable elements of the water vascular system, such as the ring canal, radial canals, and lateral canals, are not preserved. The arrangement of tube feet into two sets contradicts previous interpretations of the edrioasteroid

WVS (e.g. Bell 1977; Paul & Smith 1984; Smith 1985) and is suggestive of a system that differs radically from anything previously documented in extant or extinct echinoderms.

3.2.4 Ophiocistioidea

Ophiocistioids are extinct echinozoans characterized by a dome-shaped body with a complex jaw apparatus and long, plated tube feet. Rahman et al. (2019) described a new species, *Sollasina cthulhu* Rahman et al. 2019, from the Herefordshire Lagerstätte using physical–optical tomography and virtual reconstruction. This revealed an internal soft-tissue structure that was interpreted as the ring canal (Figure 3, parts 3–4), a key feature of the water vascular system in living echinozoans that has hitherto been undocumented in any fossil forms. No other elements of the water vascular system are preserved, but the size and arrangement of the ring canal and plated tube feet suggest the radial canals linking these structures would have been relatively short compared to extant echinoids and holothurians.

3.3 Digestive System

Reconstructing the digestive system in extinct echinoderms is challenging due to the scarcity of fossils with soft parts preserved. Most previous attempts have relied on the placement of the mouth and anus in fossils and the morphology of extant analogues. However, rare specimens with three-dimensional preservation of the gut are known (e.g. Kammer & Ausich 2007), and the study of such material can provide valuable insights into the anatomy of the echinoderm digestive system. Rahman et al. (2015) used synchrotron tomography to image an exceptionally preserved, post-metamorphic blastoid specimen from the Carboniferous of Guangxi, China. The fossil measures about one millimeter in width, necessitating the use of high-resolution synchrotron tomography to accurately resolve fine details of its anatomy. This revealed a U-shaped tubular structure within the theca, less than 200 μm in diameter, which was interpreted as the developing gut (Figure 3, parts 5–6). The U-shaped morphology of this structure differs from what is documented for the gut in comparable developmental stages of extant crinoids, suggesting the group may not be a suitable analogue for the anatomy and development of extinct stemmed echinoderms, contrary to previous suggestions (Breimer & Macurda 1972).

3.4 Nervous and Circulatory Systems

The nervous and circulatory systems of extinct echinoderms are largely inferred based on those of their extant relatives. In echinoids, among the best studied groups, the nervous system consists of a nerve ring, radial nerves, and sub-epidermal plexus, and the haemal system comprises a haemal ring and two other main channels called the conspicuous sinuses (Hyman 1955); similar structures are sometimes hypothesized to have been present in fossil forms. Under exceptional circumstances, however, fossil specimens apparently preserving elements of these systems have been reported (e.g. Haugh 1975; Saulsbury & Zamora 2020), allowing us to make comparisons with living species.

Three-dimensionally preserved specimens of the Cretaceous featherstar, *Decameros ricordeanus* d'Orbigny 1850, provided an ideal opportunity to reconstruct the nervous and circulatory anatomy with micro-CT (Saulsbury and & Zamora 2020). In these fossil specimens, internal cavities are filled with a high-density material, providing strong density contrast between the calcitic skeletal elements and the infill. Virtual fossils reveal a complex and extensive internal circulatory system. Additionally, they show that the peripheral portions of the nervous system are linked to both the circulatory system and the body wall. Based on this, Saulsbury and Zamora (2020) reconstructed a complex pattern of coelomic fluid flow for *D. ricordeanus*, which likely allowed fluid to circulate through the entire calyx coelom. This complexity would have increased the surface area of this structure, enhancing important functions such as nutrient and hormone transport and respiration. Detailed assessment of the internal anatomy of *D. ricordeanus* compared to modern featherstars suggests a high degree of variation with the body cavity size and shape, a uniform central aboral nervous system (ANS), and variation in the peripheral ANS. Understanding the implications of this variation in the peripheral ANS will require additional work on the neurobiology of extant crinoids (Saulsbury and & Zamora 2020).

Functionality of the crinoid ANS was previously investigated by Hamann (1889) and Nakano et al. (2004), but the peripheral parts of this system had not been explored in detail until the work by Saulsbury and Zamora (2020). These peripheral nerves are thought to have been well situated for sensing shifting environmental conditions owing to the placement of nerve endings on the aboral surface, and they may therefore have had a sensory function in *D. ricordeanus*.

4 Taphonomy

The taphonomy of fossil echinoderms has been a focus of study by paleontologists for many years (e.g. Lewis 1980; Smith 1990). Previous work has relied heavily on laboratory experiments (e.g. Kidwell and Baumiller 1990) to improve understanding of patterns of skeletal decay and disarticulation. However, virtual paleontology can provide additional data on important aspects such as the association between body and trace fossils and the degree of articulation and orientation of specimens within the rock, opening up new lines of enquiry in taphonomic analysis.

Stock and Veiss (2003) used micro-CT to explore the degradation of the stereom microstructure by diagenesis in Jurassic echinoids. Examination of virtual slices suggested the test fragments experienced a very different diagenetic history and environment to the spines. Specifically, the test fragments appeared to have been more greatly impacted by diagenesis, as implied by the complex variegated contrast, compared to the other skeletal elements (including spines).

Lin et al. (2010) used micro-CT to explore the effects of bioturbation on the preservation of different animal groups, including fossil echinoderms. This study focused on the Kaili Biota of the middle Cambrian of South China, with two gogiid eocrinoids in matrix and their bioturbated surroundings digitally reconstructed to assess the preservational state of the fossils. The resulting virtual trace fossils allowed for a more detailed description of burrow morphology, and showed that although the trace makers were mobile, they did not intensively bore through the echinoderm body (Lin et al. 2010).

X-ray tomography was used to analyze the preservation of Miocene *Clypeaster* Lamark 1801 specimens with fossilized epibionts (Rahman et al. 2015). One specimen of *Clypeaster* that had been heavily colonized by boring bivalves was imaged with micro-CT. This specimen was digitally reconstructed to provide detailed descriptions of trace fossils and the associated trace makers inside the echinoid, which helped identify the bivalves as members of the genus *Rocellaria* Blainville 1828. Furthermore, the virtual fossil informed on the colonization history of this unusual *Clypeaster* specimen, indicating that there were multiple episodes of postmortem colonization by boring bivalves.

Using micro-CT, Reid et al. (2019) described a well-preserved ophiuroid–stylophoran assemblage from the Lower Devonian of South Africa. Study of this assemblage using conventional methods was not feasible, as the rock had undergone severe chemical weathering and was in a fragile state. However, a virtual reconstruction shed new light on echinoderm-dominated communities of southwestern Gondwana, revealing a very high abundance of a newly described ophiuroid, *Gamiroaster tempestatis* Reid et al. 2019 (Figure 4). In addition, this enabled a detailed taphonomic analysis, including aspects such as the degree of fossil articulation, orientation, and faunal composition.

5 Phylogeny

Molecular data provide a rigorous phylogenetic hypothesis for extant deuterostomes (e.g. Mallatt and Winchell 2007; Dunn et al. 2008; Lartillot and

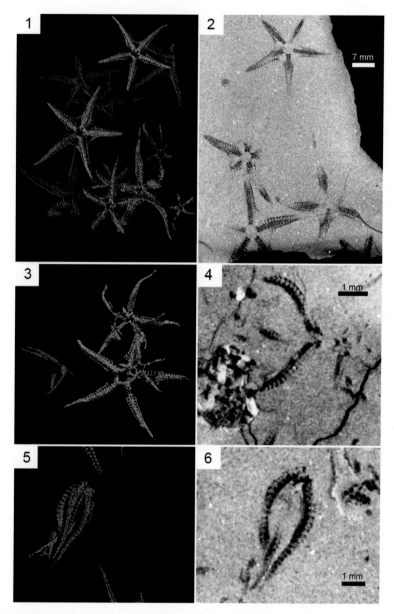

Figure 4 Example of advances in understanding of echinoderm taphonomy provided by computed tomography. (1, 3, 5) Virtual reconstructions and (2, 4, 6) micro-CT slices showing the variation in ophiuroid postures and the abundance of individuals preserved in the Karbonaatjies obrution bed. Modified from Reid et al. (2019).

Philippe 2008; Miller et al. 2017; O'Hara et al. 2017), which shows that crinoids are the sister group to the other extant classes of echinoderms (e.g., Perseke et al. 2010; Janies et al. 2011; Pisani et al. 2012). In recent years, there has been a concerted effort to incorporate fossil taxa into reconstructions of the echinoderm tree of life (e.g. Thuy and Stöhr 2016; Bauer et al. 2017, 2019; Cole 2017, 2019; Wright et al. 2017; Sheffield and Sumrall 2019a,b; Mongiardino Koch and Thompson 2020), but the relationships within and between most extinct groups are unclear or contentious (Bauer et al. 2019; Sheffield and Sumrall, 2019a,b; Sheffield et al. forthcoming in this series). Virtual paleontology will continue to shed light on this long-standing issue, for example, by enabling the description of new potentially phylogenetically informative characters.

5.1 Stem Echinoderms

All living echinoderms are characterized by pentaradial symmetry as adults, derived originally from a bilaterally symmetrical ancestral body plan (Smith 2005, 2008; Zamora and Rahman 2014). The existence of fossil echinoderms with different types of symmetry, including bilateral, asymmetrical, triradial, and pentaradial forms, could potentially shed light on this fundamental evolutionary transformation, but the phylogenetic relationships of these extinct groups are debated (David et al. 2000; Smith 2005; Sumrall and Wray 2007; Zamora and Rahman 2014). By using micro-CT to describe the bilaterally symmetrical Cambrian echinoderm *Ctenoimbricata spinosa*, Zamora et al. (2012) were able to identify morphological characters shared with ctenocystoids and cinctans. This new homology scheme informed a phylogenetic analysis of select Cambrian echinoderms by Smith and Zamora (2013), which suggested that echinoderms evolved through successive bilateral, asymmetrical, triradial, and pentaradial stages. This helped uncover the earliest steps in the assembly of the echinoderm body plan (Zamora & Rahman 2014).

5.2 Blastozoa

Among blastozoans, external expressions of the respiratory and water vascular systems have traditionally been used as synapomorphies. However, there is growing evidence that these structures are likely homoplastic and, therefore should not be used to define clades (Paul and Smith 1984; Sumrall and Gahn 2006; Sheffield and Sumrall 2019a). Methods for reconstructing the internal respiratory system of eublastoids (i.e. hydrospires) in three dimensions, as pioneered by Schmidtling and Marshall (2010) and Waters et al. (2015, 2017), enabled the generation of incredibly detailed virtual fossils. This provides a framework to begin to qualitatively describe hydrospires and quantitatively code them as characters to assess phylogenetic relatedness (Bauer et al. 2015, 2017). Incorporation of these internal anatomical characters in concert with more traditional external characters allowed for the estimation of novel relationships that were not recovered from external morphological data alone and increased the resolution of species relationships (Bauer et al. 2017).

This work exploring the phylogenetic importance of endothecal respiratory systems was extended further to assess the position of eublastoids in relation to other species within Blastoidea, as defined by Donovan and Paul (1985) to include coronoids and *Lysocystites* Miller 1889 an enigmatic Silurian echinoderm with a conservatively plated theca similar to coronoids and eublastoids. Synchrotron tomography was used to describe the external and internal anatomy of the Late Ordovician blastoid *Macurdablastus* (Bauer et al. 2019). Based on the observed differences in lancet plate morphology and internal respiratory structures (see above), *Macurdablastus* was removed from the Eublastoidea, with phylogenetic analysis placing it as sister taxon into this group (Bauer et al. 2019). This work supported previous qualitative assessments (Donovan and Paul 1985; Gil Cid et al. 1996) uniting coronoids, eublastoids, and *Lysocystites*.

Two new Cambrian blastozoans, *Dibrachicystis purujoensis* Zamora and Smith 2012 and *Vizcainoia moncaiensis* Zamora and Smith 2012, were described by Zamora and Smith (2012), the former with the aid of micro-CT. This provided detailed information on the anatomy of these fossil forms, which informed the description of a new family (Dibrachicystidae) and provided a rich character suite to estimate the evolutionary relationships of some Cambrian blastozoans (Figure 5). The inferred phylogeny suggests a high degree of homoplasy among characters, which was interpreted as due to the high degree of flexibility in the construction of the skeleton (Zamora & Smith 2012). Moreover, this suggests that the divergence of rhombiferan lineages dated back to at least the middle Cambrian. Lastly, the analysis of Zamora and Smith (2012) demonstrates that these blastozoan echinoderms possessed feeding appendages arising as extensions of the theca (i.e. arms), a character that has been described as a crinoid synapomorphy by other workers (e.g. Sprinkle et al. 2011).

5.3 Crinoidea

The internal structures preserved in the Cretaceous featherstar *Decameros ricordeanus* provided rich character data for phylogenetic inference. Micro-CT scans of select fossil specimens demonstrated that the coelom of *D. ricordeanus* is more complex than that of extant crinoids (Saulsbury &

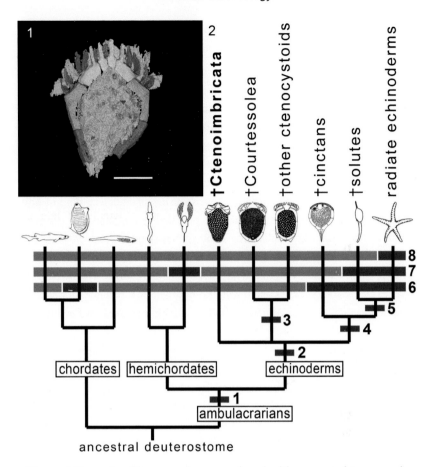

Figure 5 Example of how specimens analyzed with computed tomography provide insight into echinoderm phylogeny. (1) CT reconstructions of various elements of the stem of *Dibrachicystis purujoensis* with reference to plate locations on a specimen. (2) Evolutionary history of stalked Cambrian echinoderms with the addition of the new genus *Dibrachicystis*. Modified from Zamora and Smith (2012).

Zamora 2020), with unique configurations of coelomic morphology suggested to be clade defining. The inferred phylogeny largely agrees with previous work assessing evolutionary relationships among featherstars, with *D. ricordeanus* included in a clade characterized by the presence of coelomic diverticula. The soft tissue details derived from *D. ricordeanus* provided useful characters that resulted in a stable position in the resulting tree topology. Additional characters related to the coelomic anatomy provided support for the placement of *D. ricordeanus* with the other comatulid taxa. Virtual reconstructions by

Saulsbury and Zamora (2020) suggest that peripheral nerves are present in the radial plates of both extant and extinct crinoids.

6 Current Perspectives and Future Developments

As illustrated by the examples above, virtual paleontology has become a well-established approach for studying fossil echinoderms, providing valuable insights into their anatomy, taphonomy, and phylogeny. The rising accessibility of micro-CT technology has been integral to this growth, with an increasing number of institutions now housing scanners and software for 3D visualization. This has been accompanied by an improvement in our understanding of how best to carry out scans of fossils, with new protocols for mounting specimens, minimizing artefacts, and enhancing contrast being developed. Sharing this knowledge through publications has ensured paleontologists are better placed than ever before to study fossils with virtual reconstructions. However, until recently (Davies et al. 2017), there were no outlined best practices for sharing raw data and methodological details, both of which are essential for reproducible work.

Associated with the generation of virtual fossils is the need for online repositories to store the large raw and processed digital datasets arising from this work, which is critically important for archival purposes and to ensure access for future research (Callaway 2011; Davies et al. 2017; Sutton et al. 2017). No aggregator of virtual fossils currently exists, although models can be linked to physical specimen records through museum databases and data aggregators (e.g., iDigBio, GBIF). However, a wide range of data repositories are available, from large-scale facilities centered on national funding sources, to multidisciplinary, discipline-specific, or institutional repositories (Davies et al. 2017; Lewis 2019). These vary considerably in terms of their accessibility, cost, and versatility; researchers must select the most appropriate repository depending on the nature of their project. The move toward virtual fossils comes with many legal and copyright issues that museums and other institutions must work to address in their formal policy agreements and find a path to track usage of these digital data (Davies et al. 2017; Lewis 2019). Davies et al. (2017) put forward a set of recommendations for publishing virtual fossils, aimed at promoting reproducible, open science. Ensuring open access to data (wherever possible) is key to unlocking the potential of virtual paleontology, as this will enable further analysis in future projects (including areas that could not be anticipated at the time the data were collected), while at the same time removing some of the barriers faced by researchers (e.g. travel funds, equipment, and access to specimens). Furthermore, aggregating these data in a visible place (e.g. associated with the online *Treatise on Invertebrate Paleontology*) will serve to maximize their research usage.

Imaging fossils with weak internal density contrast, such as fossil echinoderms preserved as calcite in limestone, remains a major challenge. Techniques such as propagation-based phase-contrast X-ray tomography can provide enhanced contrast compared to conventional micro-CT; however, in some cases (e.g. the Herefordshire Lagerstätte), destructive techniques are still needed to visualize fine details of well-preserved fossil forms. Applying a combination of non-destructive techniques to the same specimen, an approach called correlative tomography, has great potential for studying fossils that are not well suited for CT. For example, Maróti et al. (2020) imaged the skeleton of a Miocene echinoid using both neutron and X-ray tomography; they combined the superior resolution offered by X-ray imaging with the enhanced contrast provided by neutron tomography (which is sensitive to certain light elements, e.g. hydrogen) to obtain a high-quality reconstruction of fossil morphology, including internal structures. Moreover, neutron tomography yields indirect information on sample chemical composition and, when combined with other elemental analysis methods (e.g. X-ray fluorescence spectroscopy), can provide a comprehensive picture of the specimen's chemistry (Maróti et al. 2020).

There remains the potential for over (or mis-) interpretation of small-scale internal features when tomographic techniques are used as a tool. In some instances, these approaches can be ground truthed with historical datasets of serial sections or acetate peels, which provide additional data on the color of structures that can aid their interpretation. Authors should explicitly address the preservational and diagenetic processes that have might altered the morphology of the specimen, as well as any potential scanning artifacts (see e.g. Sutton et al. 2014) that may be present. This is particularly important in work that uses novel morphological findings to define new species and propose clade synapomorphies.

The increasing availability of virtual fossil echinoderms provides an ideal opportunity for quantifying morphology in three dimensions and rigorously analyzing form and function. Computational methods such as range of motion analysis (e.g. Clark et al. 2020), 3D geometric morphometrics (e.g. Bright et al. 2016), computational fluid dynamics (e.g. Dynowski et al. 2016; Rahman 2020), finite element analysis (Rayfield 2007), and musculosketal modeling (e.g. Lautenschlager 2015) can be used to test functional and ecological hypotheses in enigmatic extinct species. For example, Clark et al. (2020) studied the range of motion of the appendage of an Ordovician stylophoran, demonstrating that the animal could not have moved through dorsoventral movements of the appendage, contrary to some previous suggestions. These virtual modelling approaches have been applied only rarely to fossil echinoderms thus far, but have great potential for tackling a wide range of paleobiological and evolutionary questions.

7 Conclusions

Fossil echinoderms have at long last entered the virtual world. Tomographic techniques are now widely employed and provide new avenues for exploring echinoderm anatomy non-destructively and in unprecedented detail. These methods can also shed light on other important aspects of echinoderm paleobiology, such as function, preservation, and phylogenetic relationships. Producing detailed and open workflows for applying these techniques to fossils, with modifications to parameters where necessary (e.g. scanning specimens with pyrite replacement requires different settings), will promote the accessibility and increase usage of comprehensive technical studies across Echinodermata. Ensuring the datasets and methodological details underlying this work are accessible is crucial for enabling verification and replication of published results, as well as reuse for research purposes. It would be beneficial to the community to produce an aggregated list or directory of available virtual fossil echinoderms with links to digital repositories and associated literature or files. Future work combining different imaging techniques, using novel computational methods, and ensuring full transparency with data availability and methodological details will ensure the accessible and virtual future of echinoderm paleobiology.

Supplemental Material

Synchrotron slices and scan settings for *Cryptoschisma* sp. (MGM-3384D) are available from: https://zenodo.org/record/4837028#.YRu788pKhaQ

References

Ausich, W. I., Bartels, C., & Kammer, T. W. (2013). Tube foot preservation in the Devonian crinoid *Codiacrinus* from the Lower Devonian Hunsrück Slate, Germany. *Lethaia*, **46**(3), 416–420.

Bauer, J. E., Sumrall, C. D., & Waters, J. A. (2015). In Zamora, S. and Rábano, I., eds., *Progress in Echinoderm Palaeobiology*: Cuadernos del Museo Geominero, **19**. Instituto Geológico y Minero de España, Madrid, pp. 33–36.

Bauer, J. E., Sumrall, C. D., & Waters, J. A. (2017). Hydrospire morphology and implications for blastoid phylogeny. *Journal of Paleontology*, **91**(4), 847–857.

Bauer, J. E., Waters, J. A., & Sumrall, C. D. (2019). Redescription of *Macurdablastus* and redefinition of Eublastoidea as a clade of Blastoidea (Echinodermata). *Palaeontology* **62**(6), 1003–1013.

Beaver, H. H. (1967). Morphology. In Moore, R. C., ed., *Treatise on Invertebrate Paleontology, Part S, Echinodermata 1, v. 2*: New York and Lawrence, Geological Society of America and University of Kansas, S392–S398.

Bell, B. M. (1977). Respiratory schemes in the class Edrioasteroidea. *Journal of Paleontology*, **51**(3), 619–632.

Blainville, H. M. D. (1828). In F. Cuvier et al., 1816–1830 vol. 50, *Dictionnaire des Sciences Naturelles dans lequel on Traite Méthodiquement des Différens Êtres de la Natur, Considérés soit en Eux-mêmes, d'après l'état Actuel de nos Connoissances, soit Relativement a l'Utilité qu'en Peuvent Retirer la Médecine, l'Agriculture, le Commerce et les Arts. Suivi d'une Biographie des Plus Célèbres Naturalistes*. Strasbourg and Paris: F. G. Levrault.

Branca, W. (1906). *Die Anwendung der Röntgenstrahlen in der Paläontologie*. Abhandlungen der Königlich Preussischen Akademie der Wissenschaften, Verlag der Königlichen Akademie der Wissenschaften.

Breimer, A. (1988a). The anatomy of the spiraculate blastoids; Part I: The family Troosticrinidae. *Proceedings of the Koninklijke Nederlandse Akademie van Wetenschappen, Series B*, **91**(1), 1–13.

Breimer, A. (1988b). The anatomy of spiraculate blastoids; Part II: The family Diploblastidae. *Proceedings of the Koninklijke Nederlandse Akademie van Wetenschappen, Series B*, **91**(2), 161–169.

Breimer, A., & Macurda Jr., D. B. (1972). *The Phylogeny of Fissiculate Blastoids*. Amsterdam: North-Holland Publishing Company. 390 pp.

Briggs, D. E., Siveter, D. J., & Siveter, D. J. (1996). Soft-bodied fossils from a Silurian volcaniclastic deposit. *Nature*, **382**(6588), 248–250.

Briggs, D. E. G., Siveter, D. J., Siveter, D. J., Sutton, M. D., & Rahman, I. A. (2017). An edrioasteroid from the Silurian Herefordshire Lagerstätte of England reveals the nature of the water vascular system in an extinct echino-derm. *Proceedings of the Royal Society B: Biological Sciences*, **284**(1862), 20171189. doi: https://doi.org/10.1098/rspb.2017.1189

Bright, J. A., Marugán-Lobón, J., Cobb, S. N., & Rayfield, E. J. (2016). The shapes of bird beaks are highly controlled by nondietary factors. *Proceedings of the National Academy of Sciences*, **113**(19), 5352–5357.

Broadhead, T. W. (1984). *Macurdablastus*, a Middle Ordovician blastoid from the southern Appalachians. *The University of Kansas Paleontological Contributions*, **110**, 1–9.

Callaway, E. (2011). Fossil data enter the web period. *Nature*, **472**, 150.

Clark, E. G., Bhullar, B.-A. S., Darroch, S. A. F., & Briggs, D. E. G. (2017). Water vascular system architecture in an Ordovician ophiuroid. *Biology Letters*, **13**(12), 20170635. doi: https://doi.org/10.1098/rsbl.2017.0635

Clark, E. G., Hutchinson, J. R., & Briggs, D. E. G. (2020a). Three-dimensional visualization as a tool for interpreting locomotion strategies in ophiuroids from the Devonian Hunsrück Slate. *Royal Society Open Science*, **7**(12), 201380.

Clark, E. G., Hutchinson, J. R., Bishop, P. J., & Briggs, D. E. (2020b). Arm waving in stylophoran echinoderms: Three-dimensional mobility analysis illuminates cornute locomotion. *Royal Society Open Science*, **7**(6), 200191.

Cole, S. R. (2017). Phylogeny and morphologic evolution of the Ordovician Camerata (class Crinoidea, phylum Echinodermata). *Journal of Paleontology*, **91**(4), 815–828.

Cole, S. R. (2019). Phylogeny and evolutionary history of diplobathrid crinoids (Echinodermata). *Palaeontology*, **62**(3), 357–373.

Cole, S. R., Wright, D. F., & Ausich, W. I. (2019). Phylogenetic community paleoecology of one of the earliest complex crinoid faunas (Brechin Lagerstätte Ordovician). *Palaeogeography, Palaeoclimatology, Palaeoecology*, **521**, 82–98.

Conroy, G. C., & Vannier, M. W. (1984). Noninvasive three-dimensional computer imaging of matrix-filled fossil skulls by high-resolution computed tomography. *Science*, **226**(4673), 456–458.

Cunningham, J. A., Rahman, I. A., Lautenschlager, S., Rayfield, E. J., & Donoghue, P. C. J. (2014). A virtual world of paleontology. *Trends in Ecology and Evolution*, **29**(6), 347–357.

David, B., Lefebvre, B., Mooi, R., & Parsley, R. L. (2000). Are homalozoans echinoderms? An answer from the extraxialaxial theory. *Paleobiology*, **26**(4), 529–555.

Davies, T. G., Rahman, I. A., Lautenschlager, S. et al.(2017). Open data and digital morphology. *Proceedings of the Royal Society B: Biological Sciences*, **284**(1852), 20170194.

Dean, J. D. (1999). What makes an ophiuroid? A morphological study of the problematic Ordovician stelleroid Stenaster and the palaeobiology of the earliest asteroids and ophiuroids. *Zoological Journal of the Linnean Society*, **126**(2), 225–250. doi: https://doi.org/10.1111/j.1096-3642 .1999.tb00154.x

de Lamarck, J. B. D. M. (1801). *Système des animaux sans vertèbres ou tableau général des classes, des ordres et des genres de ces animaux.* L'auteur.

Deline, B., Thompson, J. R., Smith, N. S. et al.(2020). Evolution and development at the origin of a phylum. **30**(9), 1672–1679. doi: https://doi.org/10 .1016/j.cub.2020.02.054

Dexter, T. A., Sumrall, C. D., & Mckinney, M. L. (2009). Allometric strategies for increasing respiratory surface area in the Mississippian blastoid Pentremites. *Lethaia*, **42**(2), 127–137.

Domínguez, P., Jacobson, A. G., & Jefferies, R. P. S. (2002). Paired gill slits in a fossil with a calcite skeleton. *Nature* **417**(6891): 841–844.

Donovan, S. K. & Paul, C. R. C. (1985). Coronate echinoderms from the Lower Palaeozoic of Britain. *Palaeontology*, **28**(3), 527–543.

d'Orbigny, A. (1850). *Podrome de paléontologie stratigraphique universelle des animaux mollusques & rayonnés faisant suite au cours élémentaire de paléontologie et de géologie stratigraphiques* (vol. 2). V. Masson.

d'Orbigny, A. (1840). *Histoire naturelle générale et particulière des crinoïdes vivans et fossiles, comprenant la description zoologique et géologique de ces animaux.* Chez l'auteur.

Dunlop, J. A., Wirth, S., Penney, D. et al. (2012). A minute fossil phoretic mite recovered by phase-contrast X-ray computed tomography. *Biology Letters*, **8** (3), 457–460.

Dunn, C. W., Hejnol, A., Matus, D. Q. et al. (2008). Broad phylogenomic sampling improves resolution of the animal tree of life. *Nature*, **452**(7188), 745–749.

Dynowski, J. F., Nebelsick, J. H., Klein, A., & Roth-Nebelsick, A. (2016). Computational fluid dynamics analysis of the fossil crinoid *Encrinus liliiformis* (Echinodermata: Crinoidea). *PLOS One*, **11**(5), e0156408.

Gil Cid, M. D., Domínguez Alonso, P., Cruz, M. C., & Escribano, M. (1996). Primera cita de un blastoideo Coronado en el Ordovícico Superior de Sierra

Morena Oriental. *Revista de la Sociedad geológica de España*, **9**(3–4), 253–267.

Glass A. (2006). Pyritized tube feet in a protasterid ophiuroid from the Upper Ordovician of Kentucky, USA. *Acta Palaeontologica Polonica*, **51**(1), 171–184.

Hamann, O. (1889). II. Die Crinoiden. 59–132. In *Beiträge zur Histologie der Echinodermen*, **4**. G. Fischer, Jena.

Hambach, G. (1903). *A Revision of the Blastoideae, with a Proposed New Classification, and Description of New Species*. St. Louis, MO: Nixon-Jones Printing Company.

Haubitz, B., Prokop, M., Döhring, W., Ostrom, J. H., & Wellnhofer, P. (1988). Computed tomography of *Archaeopteryx*. *Paleobiology*, **14**(2), 206–213.

Haugh, B. N. (1975). Nervous systems of Mississippian camerate crinoids. *Paleobiology*, **1**(3), 261–272.

Hendricks, J. R., Stigall, A. L., & Lieberman, B. S. (2015). *The Digital Atlas of Ancient Life*: Delivering information on paleontology and biogeography via the web. *Palaeontologia Electronica*, Article 18.2.3E.

Huynh, T. L., Evangelista, D., & Marshall, C. R. (2015). Visualizing the fluid flow through the complex skeletonized respiratory structures of a blastoid echinoderm. *Palaeontologia Electronica*, **18**(1), 1–17.

Hyman, L. H. (1955). *The Invertebrates: Echinodermata*. New York: McGraw-Hill.

Janies, D. A., Voight, J. R., & Daly, M. (2011). Echinoderm phylogeny including Xyloplax, a progenetic asteroid. *Systematic Biology*, **60**(4), 420–438.

Jefferies, R. P. S., & Lewis, D. N. (1978). The English Silurian fossil *Placocystites forbesianus* and the ancestry of the vertebrates. *Philosophical Transactions of the Royal Society of London. Series B, Biological Sciences*, **282**(990), 205–323.

Kammer, T. W., & Ausich, W. I. (2007). Soft-tissue preservation of the hind gut in a new genus of cladid crinoid from the Mississippian (Visean, Asbian) at St Andrews, Scotland. *Palaeontology*, **50**(4), 951–959.

Kidwell, S. M., & Baumiller, T. (1990). Experimental disintegration of regular echinoids: Roles of temperature, oxygen, and decay thresholds. *Paleobiology*, **16**(3), 247–271.

Kolata, D. R., Frest, T. J., & Mapes, R. H. (1991). The youngest carpoid: Occurrence, affinities, and life mode of a Pennsylvanian (Morrowan) mitrate from Oklahoma. *Journal of Paleontology*, **65**(5), 844–855.

Lartillot, N., & Philippe, H. (2008). Improvement of molecular phylogenetic inference and the phylogeny of Bilateria. *Philosophical Transactions of the Royal Society B: Biological Sciences*, **363**(1496), 1463–1472.

Lautenschlager, S. (2015). Estimating cranial musculoskeletal constraints in theropod dinosaurs. *Royal Society Open Science*, **2**(11), 150495.

Lefebvre, B., Guensburg, T. E., Martin, E. L. et al. (2019). Exceptionally preserved soft parts in fossils from the Lower Ordovician of Morocco clarify stylophoran affinities within basal deuterostomes. *Geobios*, **52**, 27–36. https://www.sciencedirect.com/science/article/pii/S0016699518301219

Lewis, D. (2019). The fight for control over virtual fossils. *Nature*, **567**(7746), 20–24.

Lewis, R. (1980). Taphonomy. *Studies in Geology, Notes for a Short Course*, **3**, 27–39.

Lin, J. P., Zhao, Y. L., Rahman, I. A., Xiao, S., & Wang, Y. (2010). Bioturbation in Burgess Shale-type Lagerstätten – Case study of trace fossil–body fossil association from the Kaili Biota (Cambrian Series 3), Guizhou, China. *Palaeogeography, Palaeoclimatology, Palaeoecology*, **292**(1–2), 245–256.

Mallatt, J., & Winchell, C. J. (2007). Ribosomal RNA genes and deuterostome phylogeny revisited. More cyclostomes, elasmobranchs, reptiles, and a brittle star. *Molecular Phylogenetics and Evolution*, **43**(1–3), 1005–1022.

Maróti, B., Polonkai, B., Szilágyi, V. et al. (2020). Joint application of structured-light optical scanning, neutron tomography and position-sensitive prompt gamma activation analysis for the non-destructive structural and compositional characterization of fossil echinoids. *NDT & E International*, **115**, 102295.

Miller, A. K., Kerr, A. M., Paulay, G. et al. (2017). Molecular phylogeny of extant Holothuroidea (Echinodermata). *Molecular Phylogenetics and Evolution*, **111**, 110–131.

Miller, S. A. (1889). *North American Geology and Palaeontology for the Use of Amateurs, Students, and Scientists*. Cincinnati, OH: Western Methodist Book Concern.

Miller, S. A., & Dyer, C. B. (1878). Contributions to palaeontology. *Journal of the Cincinnati Society of Natural History*, **1**, 24–39

Mongiardino Koch, N., & Thompson, J. R. (2020). A total-evidence dated phylogeny of echinoidea combining phylogenomic and paleontological data. *Systematic Biology*. doi: https://doi.org/10.1093/sysbio/syaa069

Moore, R. C., Lalicker, C. G., & Fischer, A. G. (1952). *Invertebrate Fossils*. New York: McGraw-Hill Book Company.

Nadhira, A., Sutton, M. D., Botting, J. P. et al. (2019). Three-dimensionally preserved soft tissues and calcareous hexactins in a Silurian sponge: Implications for early sponge evolution. *Royal Society Open Science*, **6**(7), p.190911. doi: https://doi.org/10.1098/rsos.190911.

Nakano, H., Hibino, T., Hara, Y., Oji, T., & Amemiya, S. (2004). Regrowth of the stalk of the sea lily, *Metacrinus rotundus* (Echinodermata: Crinoidea). *Journal of Experimental Zoology Part A: Comparative Experimental Biology*, **301**(6), 464–471.

Nichols D. (1972). The water-vascular system in living and fossil echinoderms. *Palaeontology*, **15**(4), 519–538.

O'Hara, T. D., Hugall, A. F., Thuy, B., Stöhr, S., & Martynov, A. V. (2017). Restructuring higher taxonomy using broad-scale phylogenomics: The living Ophiuroidea. *Molecular Phylogenetics and Evolution*, **107**, 415–430.

Orr, P. J., Briggs, D. E., Siveter, D. J., & Siveter, D. J. (2000). Three-dimensional preservation of a non-biomineralized arthropod in concretions in Silurian volcaniclastic rocks from Herefordshire, England. *Journal of the Geological Society*, **157**(1), 173–186.

Paul, C. R. C. (1967). New Ordovician Bothriocidaridae from Girvan and a reinterpretation of *Bothriocidaris Eichwald*. *Palaeontology*, **11**(4), 697–730.

Paul, C. R. C. (1968). Morphology and function of dichoporite pore-structures in cystoids. *Palaeontology*, **11**(5), 697–730.

Paul, C. R. C., & Smith, A. B. (1984). The early radiation and phylogeny of echinoderms. *Biological Reviews*, **59**(4), 443–481.

Perseke, M., Bernhard, D., Fritzsch, G. et al. (2010). Mitochondrial genome evolution in Ophiuroidea, Echinoidea, and Holothuroidea: Insights in phylogenetic relationships of Echinodermata. *Molecular Phylogenetics and Evolution*, **56**(1), 201–211.

Pisani, D., Feuda, R., Peterson, K. J., & Smith, A. B. (2012). Resolving phylogenetic signal from noise when divergence is rapid: A new look at the old problem of echinoderm class relationships. *Molecular Phylogenetics and Evolution*, **62**(1), 27–34. doi: https://doi.org/10.1016/j.ympev.2011.08.028.

Rahman, I. A. 2020. Computational fluid dynamics and its application in echinoderm paleobiology. *Elements of Paleontology*. Cambridge University Press. doi: https://doi.org/10.1017/9781108893473

Rahman, I. A., & Clausen, S. (2009). Re-evaluating the palaeobiology and affinities of the Ctenocystoidea (Echinodermata). *Journal of Systematic Palaeontology*, **7**(4), 413–426.

Rahman, I. A., & Smith, S. Y. (2014). Virtual paleontology: Computer-aided analysis of fossil form and function. *Journal of Paleontology*, **88**(4), 633–635.

Rahman, I. A., & Zamora, S. (2009). The oldest cinctan carpoid (stem-group Echinodermata), and the evolution of the water vascular system. *Zoological Journal of the Linnean Society*, **157**(2), 420–432.

Rahman, I. A., Zamora, S., & Geyer, G. (2010). The oldest stylophoran echinoderm: A new Ceratocystis from the Middle Cambrian of Germany. *Paläontologische Zeitschrift*, **84**(2), 227–237.

Rahman, I. A., Belaústegui, Z., Zamora, S., Nebelsick, J. H., Domènech, R., & Martinell, J. (2015). Miocene Clypeaster from Valencia (E Spain): Insights into the taphonomy and ichnology of bioeroded echinoids using X-ray micro-tomography. *Palaeogeography, Palaeoclimatology, Palaeoecology*, **438**, 168–179.

Rahman, I. A., Stewart, S. E., & Zamora, S. (2015). The youngest ctenocystoids from the Upper Ordovician of the United Kingdom and the evolution of the bilateral body plan in echinoderms. *Acta Palaeontologica Polonica*, **60**(1), 39–48.

Rahman, I. A., Thompson, J. R., Briggs, D. E. et al. (2019). A new ophiocistioid with soft-tissue preservation from the Silurian Herefordshire Lagerstätte, and the evolution of the holothurian body plan. *Proceedings of the Royal Society B*, **286**(1900), 20182792.

Rayfield, E. J. (2007). Finite element analysis and understanding the biomechanics and evolution of living and fossil organisms. *Annual Review of Earth and Planetary Sciences*, **35**, 541–576.

Reich, M., Sprinkle, J., Lefebvre, B. et al. (2017). The first Ordovician cyclocystoid (Echinodermata) from Gondwana and its morphology, paleoecology, taphonomy, and paleogeography. *Journal of Paleontology*, **91**(4), 735–754.

Reid, M., Taylor, W. L., Brett, C. E., Hunter, A. W., & Bordy, E. M. (2019). Taphonomy and paleoecology of an ophiuroid-stylophoran obrution deposit from the Lower Devonian Bokkeveld Group, South Africa. *Palaios*, **34**(4), 212–228.

Robison, R. A., & Sprinkle, J. (1969). Ctenocystoidea: new class of primitive echinoderms. *Science*, **166**(3912), 1512–1514.

Rowe, R., McBride, E. F., Sereno, P. C. et al. (2001). Dinosaur with a heart of stone. *Science*, **291**(5505), 783. doi: https://doi.org/10.1126/science.291.5505.783a

Salter, J. W. (1857). On some new Palaeozoic star-fishes. *Annals and Magazine of Natural History*, **20**(119), 321–334.

Saulsbury, J., & Zamora, S. (2020). The nervous and circulatory systems of a Cretaceous crinoid: preservation, palaeobiology and evolutionary significance. *Palaeontology*, **63**(2), 243–253. doi: https://doi.org/10.1111/pala.12452

Schmidtling, R. C., II, & Marshall, C. R. (2010). Three dimensional structure and fluid flow through the hydrospires of the blastoid echinoderm, *Pentremites rusticus*. *Journal of Paleontology*, **84**(1), 109–117.

Sheffield, S. L. & Sumrall, C. D. (2019a). The phylogeny of the Diploporita: a polyphyletic assemblage of blastozoan echinoderms. *Journal of Paleontology*, **93**(4), 740–752. doi: https://doi.org/10.1017/jpa.2019.2

Sheffield, S. L., & Sumrall, C.D. (2019b). A re-interpretation of the ambulacral system of *Eumorphocystis* (Blastozoa: Echinodermata) and its bearing on the evolution of early crinoids. *Palaeontology*, **62**(1), 163–173. doi: https://doi .org/10.1111/pala.12396.

Sheffield, S. L., Ausich, W. I., & Sumrall, C. D. (2018). Late Ordovician (Hirnantian) diploporitan fauna of Anticosti Island, Quebec, Canada: Implications for evolutionary and biogeographic patterns. *Canadian Journal of Earth Sciences*, **55**(1), 1–7.

Sheffield, S. L., Limbeck, M. R., Bauer, J. E., Hill, S. A., & Nohejlová, M. (forthcoming). A review of blastozoan echinoderm respiratory structures. *Elements of Paleontology*.

Siveter, D. J., Briggs, D. E., Siveter, D. J., & Sutton, M. D. (2020). The Herefordshire Lagerstätte: Fleshing out Silurian marine life. *Journal of the Geological Society*, **177**(1), 1–13. doi: https://doi.org/10.1144/jgs2019-110.

Smith, A. B. (1985). Cambrian eleutherozoan echinoderms and the early diversification of edrioasteroids. *Palaeontology* **28**(4), 715–756.

Smith, A. B. (1990). Biomineralization in echinoderms. *Skeletal Biomineralization: Patterns, Processes and Evolutionary Trends*, **1**, 413–442.

Smith, A. B. (2005). The pre-radial history of echinoderms. *Geological Journal*, **40**(3), 255–280.

Smith, A. B. (2008). Deuterostomes in a twist: The origins of a radical new body plan. *Evolution & Development*, **10**(4), 493–503.

Smith, A. B., & Zamora, S. (2013). Cambrian spiral-plated echinoderms from Gondwana reveal the earliest pentaradial body plan. *Proceedings of the Royal Society B: Biological Sciences*, **280**(1765), 20131197.

Sollas, W. J. (1904) A method for the investigation of fossils by serial sections. *Philos. Trans. R. Soc. Lond. B: Biol. Sci.* **196**, 257–263.

Sprinkle, J. (1973).*Morphology and Evolution of Blastozoan Echinoderms*. Museum of Comparative Zoology, Harvard University.

Sprinkle, J., & Wilbur, B. C. (2005). Deconstructing helicoplacoids: Reinterpreting the most enigmatic Cambrian echinoderm. *Geological Journal*, **40**(3), 281–293.

Sprinkle, J., Parsley, R. L., Zhao, Y. & Peng, J. (2011). Revision of lyracystid eocrinoids from the Middle Cambrian of South China and Western Laurentia. *Journal of Paleontology*, **85**(2), 250–255. doi: https://doi.org/10.1666/10-072.1

Stensiö, E. A. (1927). The Downtonian and Devonian vertebrates of Spitsbergen. Part I. Family Cephalaspidae. *Skrifter Svalbard Nordishavet* **12**, 1–391.

Stock, S. R., & Veis, A. (2003). Preliminary microfocus X-ray computed tomography survey of echinoid fossil microstructure. *Geological Society, London, Special Publications*, **215**(1), 225–235.

Sumrall, C. D., & Gahn, F. J. (2006). Morphological and systematic reinterpretation of two enigmatic edrioasteroids (Echinodermata) from Canada. *Canadian Journal of Earth Sciences*, **43**(4), 497–507.

Sumrall, C. D., & Waters, J. A. (2012). Universal elemental homology in glyptocystitoids, hemicosmitoids, coronoids and blastoids: Steps toward echinoderm phylogenetic reconstruction in derived blastozoa. *Journal of Paleontology*, **86**(6), 956–972.

Sumrall, C. D., & Wray, G. A. (2007). Ontogeny in the fossil record: Diversification of body plans and the evolution of "aberrant" symmetry in Paleozoic echinoderms. *Paleobiology*, **33**(1), 149–163.

Sutton, M. D., Briggs, D. E., Siveter, D. J., & Siveter, D. J. (2001). Methodologies for the visualization and reconstruction of three-dimensional fossils from the Silurian Herefordshire Lagerstätte. *Palaeontologia Electronica*, **4**(1), 1–17.

Sutton, M. D., Briggs, D. E. G., Siveter, D. J., Siveter, D. J., & Gladwell, D. J. (2005). A starfish with three-dimensionally preserved soft parts from the Silurian of England. *Proceedings of the Royal Society B: Biological Sciences*, **272**(1567), 1001–1006.

Sutton, M. D., Rahman, I. A., & Garwood, R. J. (2014). *Techniques for virtual palaeontology*. John Wiley & Sons.

Sutton, M. D., Rahman, I. A., & Garwood, R. (2017). Virtual Paleontology – an overview. *The Paleontological Society Papers*, **22**, 1–20.

Tate, J. R., & Cann, C. E. (1982). High-resolution computed tomography for the comparative study of fossil and extant bone. *American Journal of Physical Anthropology*, **58**(1), 67–73.

Thuy, B., & Stöhr, S. (2016). A new morphological phylogeny of the Ophiuroidea (Echinodermata) accords with molecular evidence and renders microfossils accessible for cladistics. *PloS One*, **11**(5)1–26.

Waters, J. A., Sumrall, C. D., White, L. E., & Nguyen, B. K. (2015). Advancing phylogenetic inference in the Blastoidea (Echinodermata). Virtual 3D reconstructions of the internal anatomy. In Zamora, S. and Rábano, I., eds., *Progress in Echinoderm Palaeobiology*: Cuadernos del Museo Geominero, **19**. Instituto Geológico y Minero de España, Madrid, 193–197.

Waters, J. A., White, L. E., Sumrall, C. D., & Nguyen, B. K. (2017). A new model of respiration in blastoid (Echinodermata) hydrospires based on CFD simulations of virtual 3D models. *Journal of Paleontology*, **91**(4), p. 662–671, doi: https://doi.org/10.1017/jpa.2017.1

Wright, D. F., Ausich, W. I., Cole, S. R., Peter, M. E., & Rhenberg, E. C. (2017). Phylogenetic taxonomy and classification of the Crinoidea (Echinodermata). *Journal of Paleontology*, **91**(4), 829–846.

Zamora, S., & Rahman, I. A. (2014). Deciphering the early evolution of echinoderms with Cambrian fossils. *Palaeontology*, **57**(6), 1105–1119. doi: https://doi.org/10.1111/pala.12138

Zamora, S., & Smith, A. B. (2012). Cambrian stalked echinoderms show unexpected plasticity of arm construction. *Proceedings of the Royal Society B: Biological Sciences*, **279**(1727), 293–298.

Zamora S., Rahman I. A., & Smith, A. B. (2012). Plated Cambrian Bilaterians Reveal the Earliest Stages of Echinoderm Evolution. *PLoS ONE*, **7**(6): e38296. doi: https://doi.org/10.1371/journal.pone.0038296

Zamora, S., Rahman, I. A., & Ausich, W. I. (2015). Palaeogeographic implications of a new iocrinid crinoid (Disparida) from the Ordovician (Darriwillian) of Morocco. *PeerJ*, **3**, e1450.

Acknowledgments

We thank Colin Sumrall for the invitation to submit this Element and for coordinating the short course that will elevate these advances in echinoderm paleobiology. I. A. R. was funded by a Museum Research Fellowship from Oxford University Museum of Natural History.

Cambridge Elements ☰

Elements of Paleontology

Editor-in-Chief
Colin D. Sumrall
University of Tennessee

About the Series
The Elements of Paleontology series is a publishing collaboration between the Paleontological Society and Cambridge University Press. The series covers the full spectrum of topics in paleontology and paleobiology, and related topics in the Earth and life sciences of interest to students and researchers of paleontology.
The Paleontological Society is an international nonprofit organization devoted exclusively to the science of paleontology: invertebrate and vertebrate paleontology, micropaleontology, and paleobotany. The Society's mission is to advance the study of the fossil record through scientific research, education, and advocacy. Its vision is to be a leading global advocate for understanding life's history and evolution. The Society has several membership categories, including regular, amateur/avocational, student, and retired. Members, representing some 40 countries, include professional paleontologists, academicians, science editors, Earth science teachers, museum specialists, undergraduate and graduate students, postdoctoral scholars, and amateur/avocational paleontologists.

Paleontological
S O C I E T Y

Cambridge Elements ≡

Elements of Paleontology

Elements in the Series

Printed in the United States
by Baker & Taylor Publisher Services